SpringerBriefs in Applied Sciences and Technology

SpringerBriefs present concise summaries of cutting-edge research and practical applications across a wide spectrum of fields. Featuring compact volumes of 50–125 pages, the series covers a range of content from professional to academic.

Typical publications can be:

- A timely report of state-of-the art methods
- An introduction to or a manual for the application of mathematical or computer techniques
- A bridge between new research results, as published in journal articles
- A snapshot of a hot or emerging topic
- An in-depth case study
- A presentation of core concepts that students must understand in order to make independent contributions

SpringerBriefs are characterized by fast, global electronic dissemination, standard publishing contracts, standardized manuscript preparation and formatting guidelines, and expedited production schedules.

On the one hand, **SpringerBriefs in Applied Sciences and Technology** are devoted to the publication of fundamentals and applications within the different classical engineering disciplines as well as in interdisciplinary fields that recently emerged between these areas. On the other hand, as the boundary separating fundamental research and applied technology is more and more dissolving, this series is particularly open to trans-disciplinary topics between fundamental science and engineering.

Indexed by EI-Compendex, SCOPUS and Springerlink.

More information about this series at http://www.springer.com/series/8884

Daejoong Kim · Kilsung Kwon ·
Deok Han Kim · Longnan Li

Energy Generation using Reverse Electrodialysis

Principles, Implementation, and Applications

 Springer

Daejoong Kim
Department of Mechanical Engineering
Sogang University
Seoul, Korea (Republic of)

Deok Han Kim
Department of Mechanical Engineering
Sogang University
Seoul, Korea (Republic of)

Kilsung Kwon
Korea Atomic Energy Research Institute
Seoul, Korea (Republic of)

Longnan Li
Department of Mechanical Engineering
Sogang University
Seoul, Korea (Republic of)

ISSN 2191-530X   ISSN 2191-5318 (electronic)
SpringerBriefs in Applied Sciences and Technology
ISBN 978-981-13-0313-5   ISBN 978-981-13-0314-2   (eBook)
https://doi.org/10.1007/978-981-13-0314-2

This Springer imprint is published by the registered company Springer Nature Singapore Pte Ltd.
The registered company address is: 152 Beach Road, #21-01/04 Gateway East, Singapore 189721, Singapore

Contents

List of Figures

List of Tables

Chapter 1
Introduction

This book comprises six main parts: evaluation of basic parameters (e.g., inlet flow rate and compartment thickness), effect on a spacer open ratio, comparison of a power generation with combinations of various resources, brine recovery in membrane-based desalination processes, a study on predesalination and chemical energy recover in reverse osmosis (RO), and investigation of ammonium bicarbonate solutions. This chapter first introduces salinity gradient energy (SGE), followed by the basics of reverse electrodialysis (RED). Also, the scope for this book is included.

1.1 Salinity Gradient Energy (SGE)

Energy is an essential factor for maintaining the quality of life and sustainable economic growth. Accordingly, energy resources to generate useful energy have been a matter of constant interest since our ancestors first learned how to make fire. The modern world is powered by fossil fuels, such as coal, oil, and natural gas, which accounted for approximately 81% of the global energy demand in 2010. However, their use leads to several unresolved issues. First, fossil fuels are a finite resource. Although our access to fossil fuels that were previously difficult to mine has been improved with new techniques, absolute reserves do not increase on a human time scale. Second, fossil fuels have an unbalanced distribution, causing threats to the national energy security of nations without their own supply of natural resources. Third, fossil fuel use causes many negative environmental effects. The emissions generated by burning fossil fuels cause air pollution, while the slag remaining after combustion can cause water pollution if not disposed of correctly. Additionally, climate change caused by carbon dioxide and other greenhouse gas emissions is threatening human survival. Therefore, alternative energy resources that would address these problems are being aggressively pursued.

D. Kim et al., *Energy Generation using Reverse Electrodialysis*,
SpringerBriefs in Applied Sciences and Technology,
https://doi.org/10.1007/978-981-13-0314-2_1

Various renewable energy resources are emerging to provide a sustainable future. Hydropower derived from falling water is an ancient renewable energy resource with many modern updates and applications. Unfortunately, it cannot completely replace fossil fuels because of its geographically limited availability and the negative environmental effects caused by damming natural watercourses. Currently, solar and wind energy are the dominant renewable energy sources. Although their output is inconsistent, various energy storage devices, including secondary batteries and redox flow batteries (RFB), have been developed to remedy the problem.

SGE, called blue energy, can be captured in river estuaries in which continuous mixing of seawater and river water occurs. Recently, SGE has attracted increasing attention owing to its high energy potential (approximately 2 TW) and stable generation. There are two competing technologies for harvesting SGE: pressure-retarded osmosis (PRO) and RED. PRO is the inverse process of reverse osmosis (RO, currently the most popular desalination process) as proposed by Sidney Loeb, a professor at Ben-Gurion University, in 1973. The driving force for PRO is the hydraulic

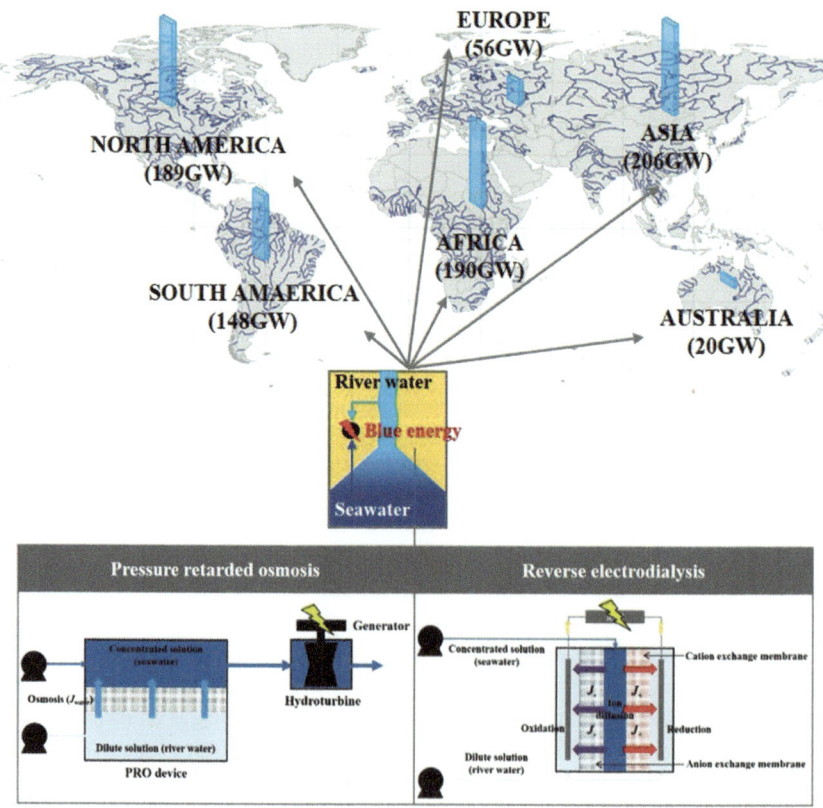

Fig. 1.1 SGE potential (top) and PRO and RED schematics (bottom)

pressure induced by water osmosis from a diluted solution (e.g., river water) to a concentrated solution (e.g., seawater), and power is generated by depressurizing the permeation through a hydropower turbine. Although RED was conceived by Pattle in the 1950s, more than 20 years before PRO, the technology was limited by difficulties in developing ion exchange membranes (IEMs) until the mid-2000s. Recently, researchers have shown renewed interest in RED systems with the rapid advancement of IEMs. While PRO and RED both have inherent advantages and disadvantages, RED typically has a smaller footprint and better reliability, making it the preferred technology. Figure 1.1 shows the SGE potential by continent and basic schematics for PRO and RED technologies.

1.2 Reverse Electrodialysis (RED)

The number of RED-related publications has exponentially increased in the last decade (Fig. 1.2), and the research scope has consistently widened. All publications before 2015 appeared in academic journals, but an increasing need for systematic knowledge of RED technology has led to a number of recent publications in edited volumes: Pawlowski et al. (2015) introduced RED technology as a type of electrokinetic phenomena and discussed important electrokinetic aspects such as IEMs, ion transport and concentration polarization, chronopotentiometry, monovalent versus multivalent ions, and redox couples. Kwon et al. (2015) summarized the fundamentals and applications of nanofluidic RED (NRED), which is a type of RED system that uses nanoporous membranes in place of conventional polymeric IEMs. Nanoporous membranes containing physical pores have the capacity to selectively transport counter-ions. The performance of an NRED is affected by the physical pore surface properties, pore size, ion mobility in working solutions, and concentration gradient. With proper scale-up and optimization, an NRED system can convert salin-

Fig. 1.2 Number of RED publications by year

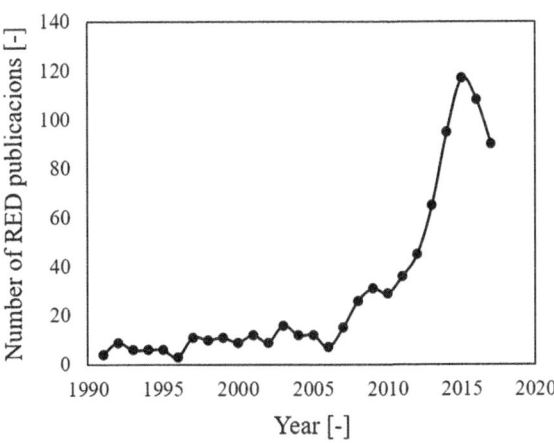

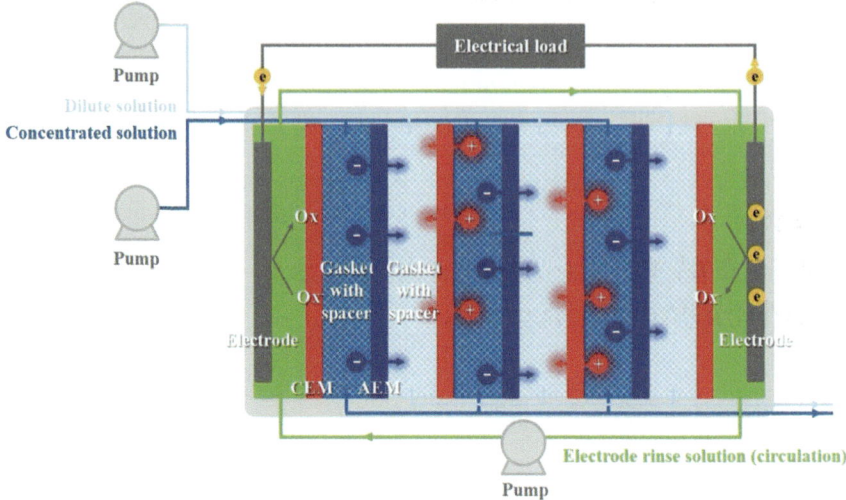

Fig. 1.3 Working principle of RED

ity energy to electrical energy. Such systems can also be scaled for utilization as a handheld power source. Another important application of NRED is as a power source in micro total analysis systems (i.e., lab on a chip). Veerman and Vermaas (2016) discussed the operating principles and fundamentals of RED technology, summarized the fundamentals and different types of SGE (e.g., PRO, RED, capacitive mixing, and mixing entropy batteries), and provided an overview of RED applications, such as the electrochemical decontamination of wastewater and microbial RED.

The working principle of RED is shown in Fig. 1.3. A mechanism for power generation is based on ion transport through the IEMs. When a concentrated solution (e.g., seawater) and a dilute solution (e.g., river water) are supplied to the system where cation exchange membranes (CEMs) and anion exchange membranes (AEMs) are alternately arranged, the ions spontaneously migrate owing to differences in chemical potential through the IEMs. Here, cations move through the CEMs and anions move through the AEMs. An electromotive force (EMF) is produced to preserve electrical neutrality in the system, and the ionic current is converted into electric current by the redox reaction on electrodes located at both ends. Electrical energy can be obtained from the movement of electrons through an external circuit in an electrical load.

The typical RED system is composed of pumps, gaskets, spacers, electrodes, and IEMs. The IEM is a key component in RED because it selectively permeates cations (CEM) or anions (AEM). Pumps are used to supply diluted and concentrated solutions and circulate the electrode rinse solution (redox couple) that reduces activation loss. Gaskets are employed for preventing leakage and spacers are located in the gaskets. Electrodes provide for electron exchange at the surface and transportation through an external circuit.

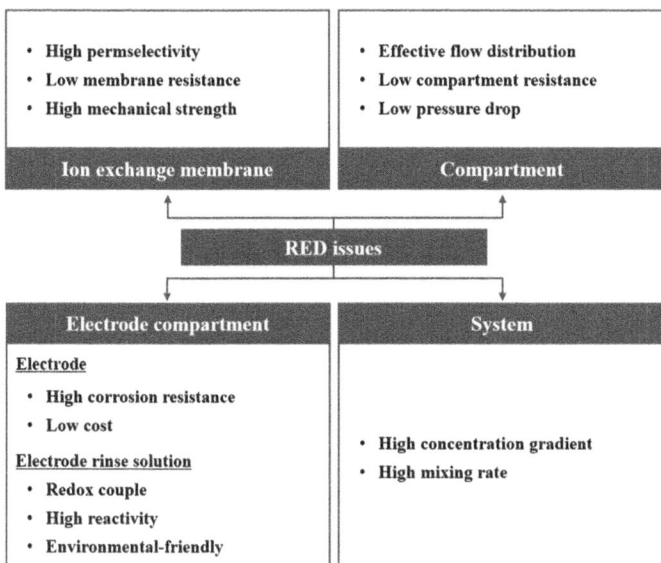

Fig. 1.4 Research issues in RED

RED technology still has some issues that prevent commercialization (Fig. 1.4). In IEMs, a reduction in membrane resistance and fabrication costs ($200 m^{-2}) is needed because the required permselectivity value is high (above 0.9) for most commercial membranes. There is also a need to improve flow geometries to achieve low pressure drop, low compartment resistance, and even flow distribution. Durability and high activity are needed in electrode compartments, including electrode strips and the rinse solution. From a system-wide viewpoint, increasing the concentration difference and mixing rate is critical for improving performance.

The IEM is a key component in RED that selectively separates cations and anions. It typically consists of hydrophobic substrates, fixed ion-functionalized groups, and movable counter-ions (Ran et al. 2017). The ion polarity passing through an IEM is determined according to the ion-functionalized groups. Researchers applied commercial IEMs manufactured by Tokuyama Corporation (Japan), Asahi Glass Corporation (Japan), FumaTech GmbH (Germany), and Hangzhou Qiangqiu Industry Corporation (China) in the early development of RED (Długołecki et al. 2008; Veerman et al. 2009a, b, 2011a, b). Veerman et al. (2009b) compared six commercial IEMs in equivalent conditions (active area of 100 cm^2 and channel thickness of 200 µm); they included one heterogeneous IEM and five homogeneous IEMs. The lowest power density was obtained for the heterogenous membrane, while a maximum power density of 1.2 W m^{-2} was obtained for the Fumasep FAD/FKD and Selemion AMV/CMV membranes (Fig. 1.4). Güler et al. (2013) measured six membrane characteristics (thickness, resistance, permselectivity, swelling degree, ion exchange capacity, and fixed charge density) associated with ion transport and power density. They used five commercial membranes, including the FumaTech FKS/FAS, FKD/FAD, Astom CMX/AMX, Selemion CMV/AMV, and Ralex CMH/AMH. The experimental condition was fixed with a flow thickness of 200 μm, temperature of 25 °C, dilute solution of 0.017 M, and concentrated solution of 0.513 M. The highest power density (1.2 W m^{-2}) was obtained in the FumaTech FAD/FKD (Fig. 1.4), an identical result to that of Veerman et al. (2009b).

Conventional commercial IEMs were not optimized for RED and have a high cost. Consequently, there remained a need for a tailor-made IEM. Güler et al. (2012) developed tailor-made AEMs composed of polyepichlorohydrin (PECH) and an amine functional group. They evaluated the membrane properties with a change in the molar ratio of the amine component and compared five different customized AEMs with the Neosepta AMX, while fixing the CEM (Neosepta CMX). Higher power density was demonstrated for customized IEMs with smaller thicknesses (33 and 77 μm). Güler et al. (2013) developed customized CEMs consisting of sulfonated polyetheretherketone (SPEEK) and a sulfonic acid functional group. The maximum power density (1.28 W m^{-2}) was found for the customized CEM and AEM used together. Lee et al. (2017) used the customized IEMs prepared by the pore-filling method. These IEMs had a significantly lower membrane resistance compared with the conventional IEMs owing to thinner membrane thicknesses, which ranged between 15 and 20 μm. These membranes achieved up to 2.43 W m^{-2} of power density, which was 1.3–1.8 times higher than that obtained from any commercial membrane pair.

The structure of the flow channel is a main factor affecting not only RED performance but also pumping loss; as such, it has gathered much attention. A flow channel has various parameters including shape, thickness, and length. Generally, a woven mesh structure is inserted in the internal flow channel of an RED, which is then called a spacer-filled RED. The mesh spacer plays important roles in distributing the flow uniformly, enhancing mixing, and supporting the membrane (Schwinge et al. 2002a, b; Gurreri et al. 2015, 2016). On the other hand, there are two negative effects from using a spacer: a shadow effect that reduces the ion exchange area, and an increase in flow resistance. Vermaas et al. (2011a) reported a power density of 2.2 W m^{-2} (channel thickness of 0.1 mm). The net power density trend showed a dramatic decrease as the channel thickness became very thin because of the greater pressure drop. Accordingly, the use of an ultrathin channel is impractical for RED. Furthermore, thin channels tend to be more vulnerable to fouling. Długołecki et al. (2010) used an ion-conductive spacer to reduce the spacer shadow effect and reported a power enhancement. An interesting concept of profiled membranes in which the flow channel is formed directly has been proposed. This concept was first demonstrated by Vermaas et al. (2011b) and has been advanced and improved by several researchers (Güler et al. 2014; Gurreri et al. 2015; Pawlowski et al. 2017).

1.3 Scope

The main body of this book is organized into five chapters: The first two chapters are parametric studies on RED performance; the remaining chapters present the analyses of RED-hybrid systems, including brine recovery in membrane-based desalination processes and low-grade waste heat recovery.

References

P. Długołecki, K. Nijmeijer, S. Metz, M. Wessling, Current status of ion exchange membranes for power generation from salinity gradients. J. Membr. Sci. **319**, 214–222 (2008)

P. Długołecki, J. Dabrowska, K. Nijmeijer, M. Wessling, Ion conductive spacers for increased power generation in reverse electrodialysis. J. Membr. Sci. **347**, 101–107 (2010)

E. Güler, Y. Zhang, M. Saakes, K. Nijmeijer, Tailor-made anion exchange membranes for salinity gradient power generation using reverse electrodialysis. Chemsuschem **5**, 2262–2270 (2012)

E. Güler, R. Elizen, D.A. Vermaas, M. Saakes, K. Nijmeijer, Performance-determining membrane properties in reverse electrodialysis. J. Membr. Sci. **446**, 266–276 (2013)

E. Güler, R. Elizen, M. Saakes, K. Nijmeijer, Micro-structured membranes for electricity generation by reverse electrodialysis. J. Membr. Sci. **458**, 136–148 (2014)

L. Gurreri, M. Ciofalo, A. Cipollina, A. Tamburini, W. van Baak, G. Micale, CFD modelling of profiled-membrane channels for reverse electrodialysis. Desalin. Water Treat. **55**, 3404–3423 (2015)

L. Gurreri, A. Tamburini, A. Cipollina, M. Micale, M. Ciofalo, Flow and mass transfer in spacer-filled channels for reverse electrodialysis: a CFD parametric study. J. Membr. Sci. **497**, 300–317 (2016)

K. Kwon, J. Han, B.H. Park, Y. Shin, D. Kim, Brine recovery using reverse electrodialysis in membrane-based desalination processes. Desalination **362**, 1–10 (2015)

M.S. Lee, H.K. Kim, C.S. Kim, H.Y. Suh, K.S. Nahm, Y.W. Choi, Thin pore filled ion exchange membranes for high power density in reverse electrodialysis: effects of structure on resistance, stability, and ion selectivity. Chemistry **2**, 1974–1978 (2017)

S. Pawlowski, J. Crespo, S. Velizarov, (2015) Sustainable power generation from salinity gradient energy by reverse electrodialysis, in *Electrokinetics Across Disciplines and Continents: New Strategies for Sustainable Development*, ed. by A.B. Ribeiro, E.P. Mateus, N. Couto (Springer International Publishing, 2015)

S. Pawlowski, T. Rijnaarts, M. Saakes, K. Nijmeijer, J.G. Crespo, S. Velizarov, Improved fluid mixing and power density in reverse electrodialysis stacks with chevron-profiled membranes. J. Membr. Sci. **531**, 111–121 (2017)

J. Ran, L. Wu, Z. Yang, Y. Wang, C. Jiang, L. Ge, E. Bakangura, T. Xu, Ion exchange membranes: new development and applications. J. Membr. Sci. **522**, 267–291 (2017)

J. Schwinge, D.E. Wiley, D.F. Fletcher, Simulation of the flow around spacer filaments between narrow channel walls. 1. Hydrodynamics. Ind. Eng. Chem. Res. **41**, 2977–2987 (2002a)

J. Schwinge, D.E. Wiley, D.F. Fletcher, Simulation of the flow around spacer filaments between channel walls. 2. Mass-transfer enhancement. Ind. Eng. Chem. Res. **41**, 4879–4888 (2002b)

J. Veerman, D.A. Vermas, Reverse electrodialysis: fundamentals, in *Sustainable Energy from Salinity Gradients*, ed. by A. Cipollina, G. Micale (Woodhead Publishing, 2016)

J. Veerman, M. Saakes, S.J. Metz, G.J. Harmsen, Reverse electrodialysis: performance of a stack with 50 cells on the mixing of sea and river water. J. Membr. Sci. **327**, 136–144 (2009a)

J. Veerman, R.M. de Jong, M. Saakes, S.J. Metz, G.J. Harmsen, Reverse electrodialysis: comparison of six commercial membrane pairs on the thermodynamic efficiency and power density. J. Membr. Sci. **343**, 7–15 (2009b)

D.A. Vermaas, M. Saakes, K. Nijimeijer, Doubled power density from salinity gradient at reduced intermembrane distance. Environ. Sci. Technol. **45**, 7089–7095 (2011a)

D.A. Vermaas, M. Saakes, K. Nijimeijer, Power generation profiled membranes in reverse electrodialysis. J. Membr. Sci. **385–386**, 234–242 (2011b)

Chapter 2
Parametric Study on RED with Sodium Chloride Solution

Here, we describe the experimental evaluation of RED performance with several parameters, including the inlet flow rates of the concentrated solution, dilute solution, and the electrode rinse solution, the concentration difference, and the intermembrane distance.

Figure 2.1 shows a schematic of the experimental setup. The RED was composed of endplates, electrodes, spacers, an anion exchange membrane, and a cation exchange membrane. The endplates were made by Poly(methyl methacrylate) (PMMA) and included the electrodes, flow field for the electrode rinse solution, and path to supply a concentrated and dilute solution. Titanium coated by iridium and ruthenium was used as the electrode. An electrode rinse solution based on a hexacyanoferrate system was used to reduce the power loss generated by the conversion process from ionic current to electron current. Peristaltic pumps were used to supply the working solutions in each compartment. The I–V characteristic (i.e., the polarization curve) was measured using a Sourcemeter (Keithley 2410). The concentrated and diluted solutions were prepared on the basis of total dissolved solids (TDS) of seawater (~35,000 TDS) and river water (~600 TDS) and were ~0.6 and ~0.01 mol/L, respectively.

Figure 2.2 shows the gross power density, pumping power density, and net power density with the inlet flow rate ranging from 1 to 20 mL/min. The black line, dotted line, and red line represent the gross power density, pumping power density, and net power density, respectively. The gross power density increases as the flow rate increases because a similar concentration difference is maintained for the whole active membrane area. The pumping power has a quadratic function of the flow rate. Accordingly, the net power density shows a non-monotonic dependence with the flow rate. The maximum point of net power density, in this case, was 7.5 mL/min.

© The Author(s), under exclusive license to Springer Nature Singapore Pte Ltd. 2019
D. Kim et al., *Energy Generation using Reverse Electrodialysis*,
SpringerBriefs in Applied Sciences and Technology,
https://doi.org/10.1007/978-981-13-0314-2_2

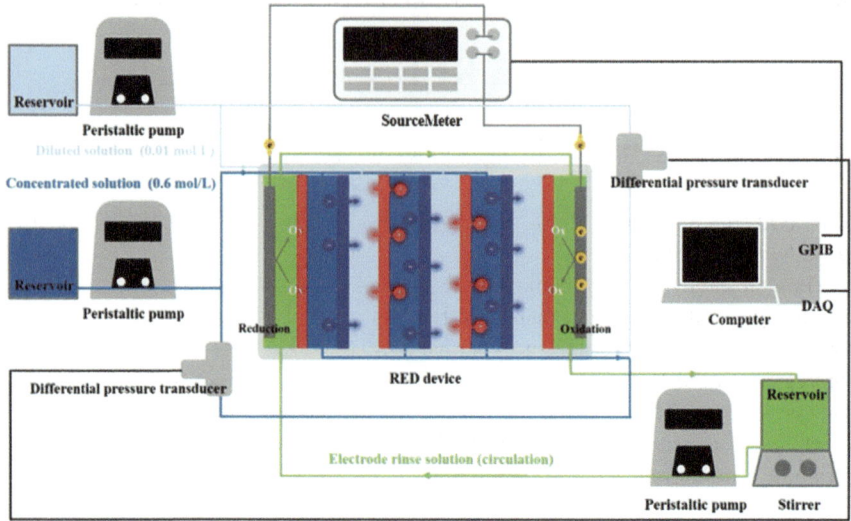

Fig. 2.1 Schematic of the experimental setup

Fig. 2.2 Power density as a function of flow rate

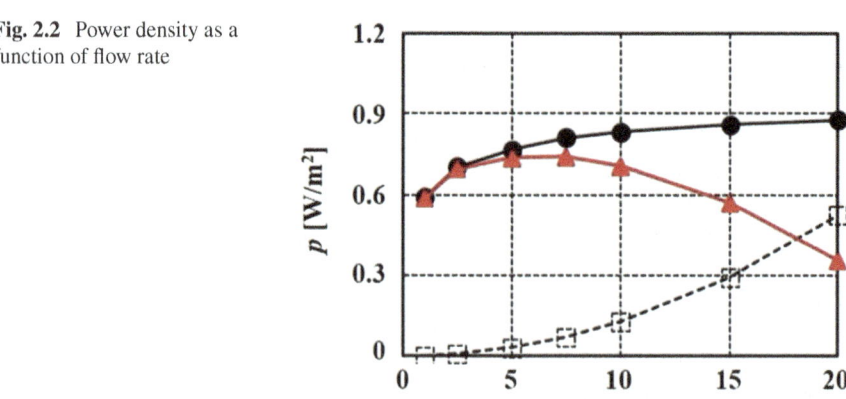

We performed the test with variations in the flow rate of the electrode rinse solution. The flow rate of the diluted and concentrated solution was fixed at 10 mL/min. Figure 2.3 shows the maximum power density as a function of the flow rate of the electrode rinse solution; we observed a monotonic increase in maximum power density when the flow rate of the electrode rinse solution increased.

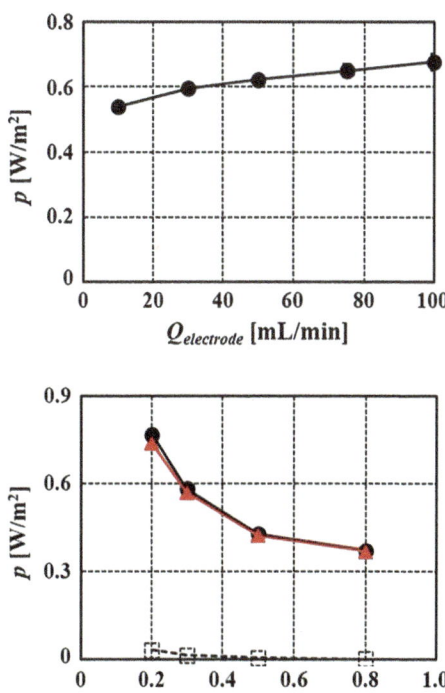

Fig. 2.3 Maximum power density as a function of electrode rinse solution flow rate: $C_H = 0.6$ mol/L, $C_L = 0.01$ mol/L, $Q_{cell} = 10$ mL/min, $\delta = 0.3$ mm

Fig. 2.4 Maximum power density as a function of intermembrane distance: $C_H = 0.6$ mol/L, $C_L = 0.01$ mol/L, $Q_{cell} = 5$ mL/min, $Q_{electrode} = 50$ mL/min, $\delta = 0.2$ mm

The intermembrane distance is associated with the internal resistance (Vermaas et al. 2011a). We measured the power density in RED with four different intermembrane distances (0.2, 0.3, 0.5, and 0.8 mm). Figure 2.4 shows the maximum (black line), pumping (dot line), and net power (red line) densities as a function of intermembrane distance. The maximum power density significantly increases with decreasing intermembrane distance owing to the reduction in the compartment resistance (especially, the diluted compartment). The pumping power increases as the intermembrane distance decreases. In that case, the net power density is maximized at the intermembrane distance of 0.2 mm. When the flow rate increases, this trend can be varied.

Figure 2.5 shows the open circuit voltage (OCV) and the power density with changes in the diluted resources; for example, changes in river water (~0.01 mol/L), wastewater (~0.05 mol/L), and brackish water (~0.085 mol/L) at a fixed concentrated solution (seawater ~0.6 mol/L). The OCV and the maximum power density decreased with increasing the concentration of the diluted solution.

This chapter presented an evaluation of several parameters that affect RED performance. Power density increases with increasing flow rate of the concentrated, diluted, and electrode rinse solutions and decreasing intermembrane distance. The pumping power consumption depends on the flow rate and the intermembrane distance. The

Fig. 2.5 **a** Open circuit
voltage and **b** maximum
power density as a function
of the diluted solution
including river water
(~0.01 mol/L), wastewater
(~0.05 mol/L), and brackish
water (~0.085 mol/L): C_H =
0.6 mol/L, Q_{cell} =
10 mL/min, $Q_{electrode}$ =
50 mL/min, δ = 0.2 mm, and
N_{cell} = 5

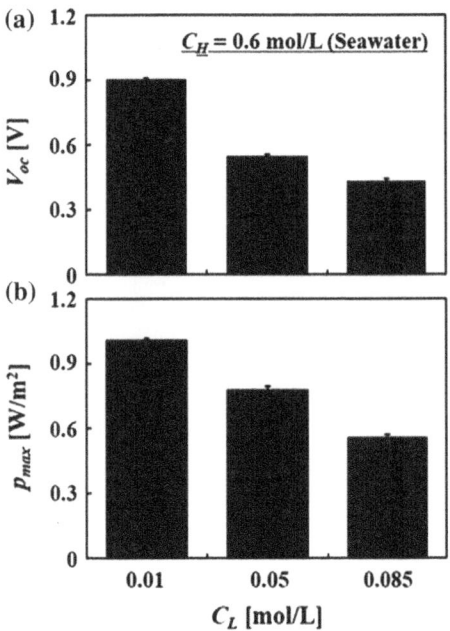

net power density has a maximum value for an inlet flow rate of 7.5 mL/min and an
intermembrane distance of 0.2 mm.

Chapter 3
Effect of Flow Structure on RED Performance

3.1 Power Enhancement of RED with a Highly Opening Spacer

We evaluated RED performance using two polymer-woven spacers with open ratios of 37 and 56% under identical experimental conditions. The maximum voltage and maximum power were compared with the inlet flow rate. The effects of ohmic and the nonohmic resistances were analyzed using the current-interrupt method for each spacer. We also investigated variation in the pressure drop to compute the pumping losses. Finally, a comparison of net power was performed for the 37 and 56% spacers.

We used the plate-flame-type RED stack, which is a serial series of a unit cell, for the experiment. The RED stack was composed of electrodes, endplates, CEMs, AEMs, gaskets, and spacers. There were three cell pairs, and an additional CEM was included to prevent the electrode rinse solution from escaping into the working solutions. The electrodes were made of titanium mesh (7 cm × 7 cm in size) coated with a one-to-one mixture of iridium and ruthenium. They were located at the PMMA endplates, which had a thickness of 30 mm. The PMMA endplates had flow paths to supply the working solutions. We employed Neosepta® CMX and AMX (Tokuyama Co., Japan) as the CEM and AEM, respectively. Silicone rubber with a thickness of 0.3 mm was used as the gasket to block internal and external leakages, which can affect performance. The flow channel to insert the spacer was formed by a laser cutting into the silicone gasket. The gasket had narrow branches to connect the inlet and outlet, and an active area of 7 cm × 7 cm to migrate ions. We applied the polymer-woven meshes (Nitex®, SEFAR, Switzerland), which had a different opening ratio. Table 3.1 presents the specifications of the two different spacers. The spacers were cut into the channel shape using the laser cutting machine. We assembled the RED stack using a digital torque wrench.

A polarization curve, which shows stack voltage against current density, is the most common method of representing the performance of electrochemical cells. We first obtained the polarization curve to compare the impact of the two different spacers

© The Author(s), under exclusive license to Springer Nature Singapore Pte Ltd. 2019
D. Kim et al., *Energy Generation using Reverse Electrodialysis*,
SpringerBriefs in Applied Sciences and Technology,
https://doi.org/10.1007/978-981-13-0314-2_3

Table 3.1 Specification of spacers

	Nitex® 06-240/37	Nitex® 06-475/56
Open ratio (%)	37	56
Mesh opening (μm)	240	475
Thickness (μm)	290	290
Wire diameter (μm)	155	180

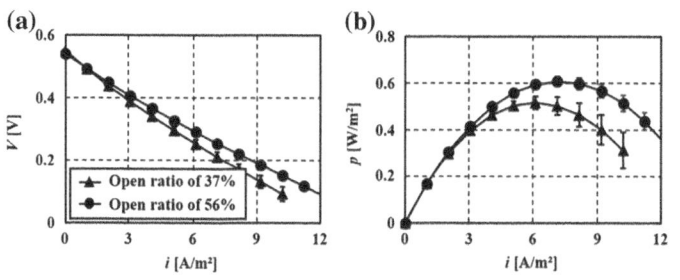

Fig. 3.1 **a** Polarization curve and **b** power density curve for two different spacers at a fixed unit cell flow rate of 6 mL/min. The triangle (▲) and the circle ((●)) symbols depict open ratios of 37% and 56%, respectively

on RED. The polarization curve was measured in the galvanostatic mode where the current is swept. The current was varied by steps of 10 mA/min. We repeated the experiment three times or more for each condition and plotted the polarization curve with error bars. The error was computed as $x = (\bar{x}+t\sigma)/\sqrt{N}$, where $\bar{x}$ is the average value, t is the t-distribution value, σ is the standard deviation, and N is the number of experiments.

Figure 3.1a presents the polarization curve of RED with the two different spacers. The unit cell flow rate was fixed at 6 mL/min. The triangle (▲) and the circle ((●)) denote the results of the 37% open ratio and 56% open ratio, respectively. The maximum voltage (i.e., the OCV) can be obtained at zero current, which means non-connecting an external circuit between the anode and the cathode. There is little difference between the spacer open ratio of 37% (0.548 V) and 56% (0.539 V). The polarization curve shows a linear trend. We found that the total resistance increases with a decreasing open ratio because the I–V slope of the 37% open ratio was steeper than that of the 57% open ratio.

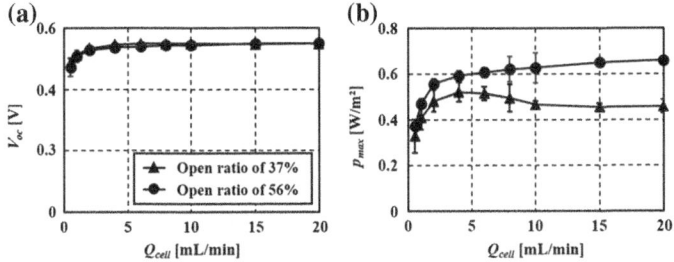

Fig. 3.2 **a** Open circuit voltage and **b** maximum power density as a function of the unit cell flow rate ranging from 0.5 to 20 mL/min in unequal intervals

The power density can be obtained from the simple formula $p = V \cdot i$ with p = power density (W/m^2) and i is the current density (A/m^2). Figure 3.1b shows the power density curve of RED computed from the polarization curve. The power density curve has a parabolic shape owing to the linear variation of the i-V curve. We accordingly obtained the maximum power density at the center of the maximum current where the external resistance equals the internal resistance. The maximum power density of the 37% open ratio (0.52 W/m^2) was found to be lower than that of the 57% open ratio (0.60 W/m^2).

Figure 3.2a shows the experimentally measured OCV as a function of the flow rate per unit cell ranging between 0.5 and 20 mL/min in unequal intervals. The flow rate significantly affected the OCV. In particular, at the low flow rate (<4 mL/min), the OCV decreases owing to the increase in the influence of the concentration boundary formed at the interface between the IEM and the working solution as well as the decrease in the concentration difference induced by the rise in the residence time. The incremental rate of the OCV decreased for flow rates exceeding 6 mL/min and the OCV gradually converged towards the maximum value. Overall, there was less than 3% difference in the OCV between the 37% open ratio and 56% open ratio.

The maximum power density as a function of the flow rate is shown in Fig. 3.2b. The maximum power density showed a monotonous increase with increasing flow rate for the spacer open ratio of 56%. Consequently, we obtained a maximum power density of 0.66 W/m^2 at a flow rate of 20 mL/min. The profile of the maximum power density has a different trend for the spacer open ratio of 37%; the maximum power density dramatically increased when the inlet flow rate was less than 4 mL/min and then monotonously decreased until 10 mL/min. It was nearly constant between 10 and 20 mL/min. The maximum power density was 0.52 W/m^2 at an inlet flow rate of 4 mL/min.

We performed quantitative analysis on the internal resistance using the current-interrupt method (i.e., chronopotentiometry), which is an electrochemical technique. A current of 80 mA where the maximum power was obtained was applied for every flow rate using the SourceMeter. The durations for each flow rate were different to ensure that equilibrium was reached. Figure 3.3 shows the variations in the current and the voltage with time at the fixed flow rate of 6 mL/min. When the current is

Fig. 3.3 Current-interrupt method to separate ohmic and nonohmic effects in the RED system with a unit cell flow rate of 6 mL/min

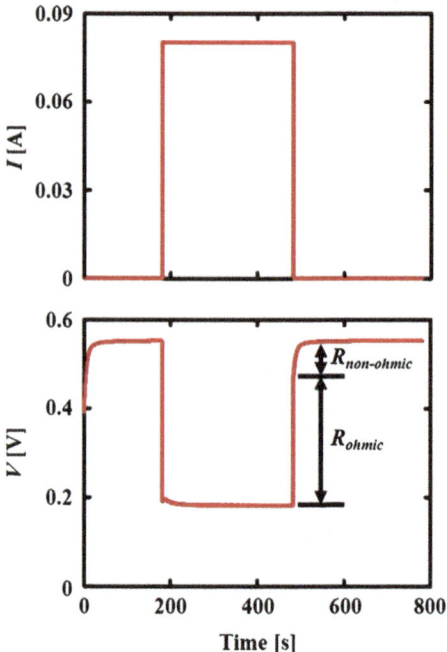

switched from the 80 mA to 0 mA, the potential jumps vertically and then gradually reaches a specific potential. Here, the potential difference caused by the vertical jump represents the ohmic effect. The nonohmic influence is the difference between the jumped potential and the plateaued potential [45].

Figure 3.4 shows the internal resistance calculated by the chronopotentiometry with two different spacers as a function of the inlet flow rate. The unit cell flow rates were 0.5, 1, 2, 4, 6, 8, 10, 15, and 20 mL/min. The internal resistance in RED can be mostly divided into ohmic resistance and nonohmic resistance. The ohmic resistance includes the membrane and the solution resistances. The nonohmic resistance is due to the concentration variation and diffusion boundary layer (DBL) formation along the flow direction. The nonohmic resistance is the difference between the total resistance and the ohmic resistance. For the spacer open ratio of 37%, the ohmic resistance monotonously increased and the nonohmic resistance decreased with increasing flow rate. In contrast, the total resistance showed a non-monotonic dependence. This is due to the dramatic reduction in the nonohmic resistance at low flow rate regimes (<2 mL/min) and the higher growth rate of ohmic resistance as compared with the decrement of the nonohmic resistance for flow rates of more than 4 mL/min. The lowest resistance (4.5 Ω) occurred at a flow rate of 2 mL/min as the polymer-woven mesh with the 37% open ratio was utilized as the spacer in the RED system. We believe that the non-monotonic profiles of the internal resistance result in the trend of the maximum power density at RED with the 37% open ratio spacer. We initially expected that the RED system with the higher opening spacer would show a

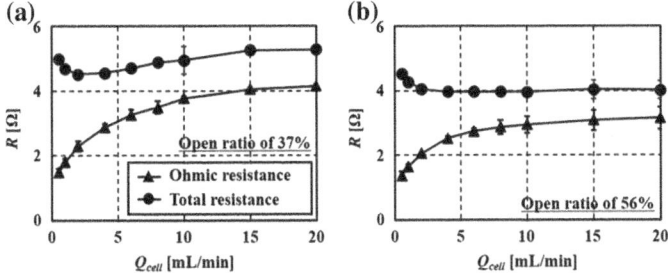

Fig. 3.4 Influences of ohmic and the nonohmic resistances with variations in the unit cell flow rate: **a** spacer open ratio of 37% and **b** spacer open ratio of 56%

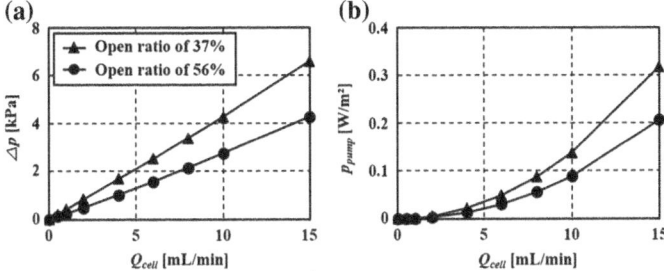

Fig. 3.5 Comparison of **a** the pressure drop and **b** pumping power consumption between two different spacer open ratios

lower ohmic resistance because of the reduction in the shadow effect, which blocks ion transport by the nonconducting region. The experimental results agree with our expectation. The ohmic resistance of the 56% open ratio was significantly lower than that of 37% open ratio for the whole range of the inlet flow rate. However, there was little difference in nonohmic resistance. The total resistance showed a decreasing trend for the spacer with the open ratio of 56%.

We measured the pressure drop with two different spacers using the differential pressure transducer and with the flow rate ranging between 0 and 15 mL/min. The pressure drops of RED systems with a spacer are typically several times higher than those of RED systems without a spacer owing to the high flow resistance (Vermaas et al. 2011). The pressure drop shows a linear increase with increasing flow rate, as shown in Fig. 3.5a. We found that flow resistance was lower for the higher open ratio. The pressure drops with the 37 and 56% open ratios were 8.1 and 5.3 kPa, respectively, at a flow rate of 15 mL/min. We calculated the pumping power by multiplying the inlet flow rate and the pressure drop (Fig. 3.5b). As such, the pumping power has a quadratic function of the inlet flow rate. The pumping power using the spacer open ratio of 37% was at least 50% higher than that using the spacer open ratio of 56%. We obtained pumping powers of 0.32 W/m^2 (spacer open ratio of 37%) and 0.21 W/m^2 (the open ratio of 56%) at a flow rate of 15 mL/min.

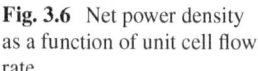

Fig. 3.6 Net power density
as a function of unit cell flow
rate

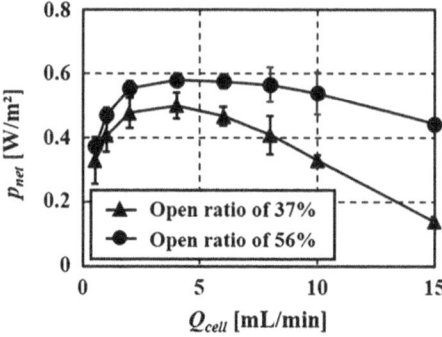

Figure 3.6 shows the net power density according to the variation in the flow rate. The net power density was computed by the power density generated in the RED system minus the consumed pumping power density. The net power density was nearly maximized at a flow rate of ~4 mL/min for the spacer open ratio of 37%, and ~6 mL/min for the spacer open ratio of 56%. We observed a rapid decrease in the net power density with the spacer open ratio of 37% because of the reduction in the gross power and the higher pumping power. The maximum net power densities were 0.52 and 0.61 W/m^2 with the 37% open ratio and the 56% open ratio, respectively.

3.2 Comparison of Spacer-Filled and Spacerless Compartments

A typical channel in RED includes polymer-woven meshes to block compartments, stabilize the flow distribution, and decrease the effect of the concentration polarization near the interface between solutions and membranes. However, spacer-filled compartments result in side effects, including the shadow effect induced by the obstruction of the ion movement and the high pumping cost. Table 3.2 presents the advantages and disadvantages of the spacer-filled channel and spacerless channel. Here, we compared the RED performance with two different geometries.

We compared the OCV and the maximum power density with two different channel designs along with a flow rate ranging between 1 and 20 mL/min (Fig. 3.7). The OCV and the maximum power density increased with increasing inlet flow rate. The OCV values with the spacerless channel were lower than those with the spacer-filled channel. In particular, for low flow rate regimes, there was a huge difference in the maximum power density of the spacer-filled channel and the spacerless channel; this was reduced when the flow rate increased.

Figure 3.8 shows the internal resistance calculated by the current-interrupt method with two different channel geometries as a function of the inlet flow rate. The ohmic resistance for the spacer-filled channel is higher than that for the spacerless channel. The nonohmic resistance monotonously decreases with the inlet flow rate. For the

Table 3.2 Advantages and disadvantages of spacer-filled and spacerless channels

Spacer-filled design	Spacerless design
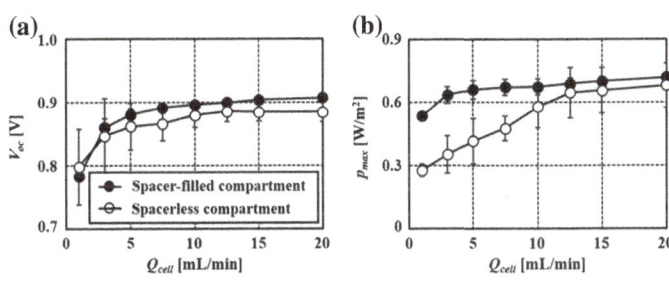	
Advantages	**Advantages**
• Weak concentration polarization	• No shadow effect
• Uniform flow distribution	• Low pressure drop
• Prevention of channel blockage	
Disadvantages	**Disadvantages**
• Shadow effect	• Strong concentration polarization
• High pressure drop	• Uneven flow distribution

Fig. 3.7 **a** Open circuit voltage and **b** maximum power density as a function of the unit cell flow rate ranging from 1 to 20 mL/min in unequal intervals

spacerless channel, the nonohmic effect is high owing to concentration polarization. Profiles of total resistance are nearly constant for the spacer-filled compartment but exhibit a decreasing trend for the spacerless compartment.

We calculated the pumping power by measuring the pressure drop, as shown in Fig. 3.9. The pumping power consumption of the spacer-filled channel was five times higher than that of the spacerless channel.

Figure 3.10 shows the net power density according to the variation in the flow rate. The maximum values of the spacer-filled and spacerless channels are similar, while their maximum points are different.

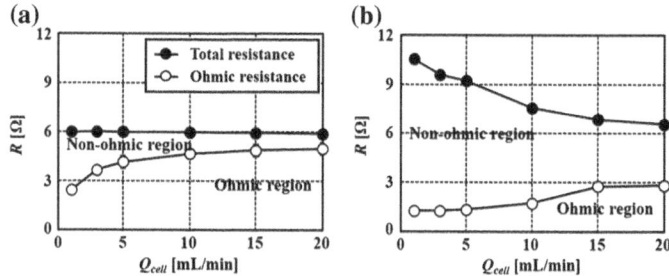

Fig. 3.8 Influences of ohmic and nonohmic resistances with variations in the unit cell flow rate: **a** spacer-filled channel and **b** spacerless channel

Fig. 3.9 Pumping power consumption as a function of the unit cell flow rate for spacer-filled and spacerless channels

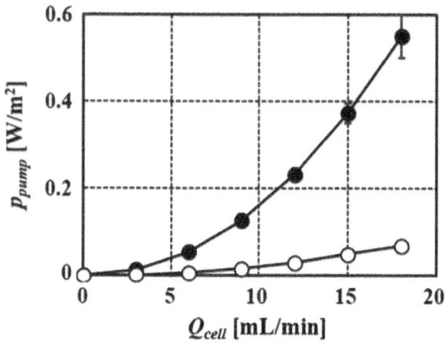

Fig. 3.10 Net power density as a function of the unit cell flow rate for spacer-filled and spacerless channels

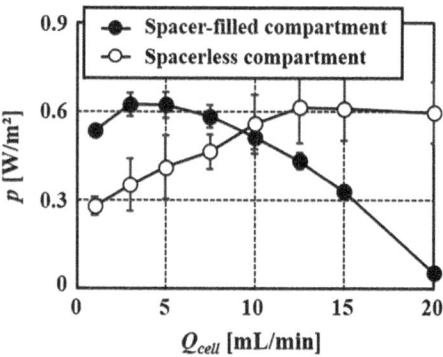

3.3 Evaluation of a Defective Spacer

We verified the importance of the high open ratio; however, the open ratio of polymer-woven meshes used as the spacer can be limited. We proposed possible designs for defective spacers to ensure a high open ratio (Fig. 3.11).

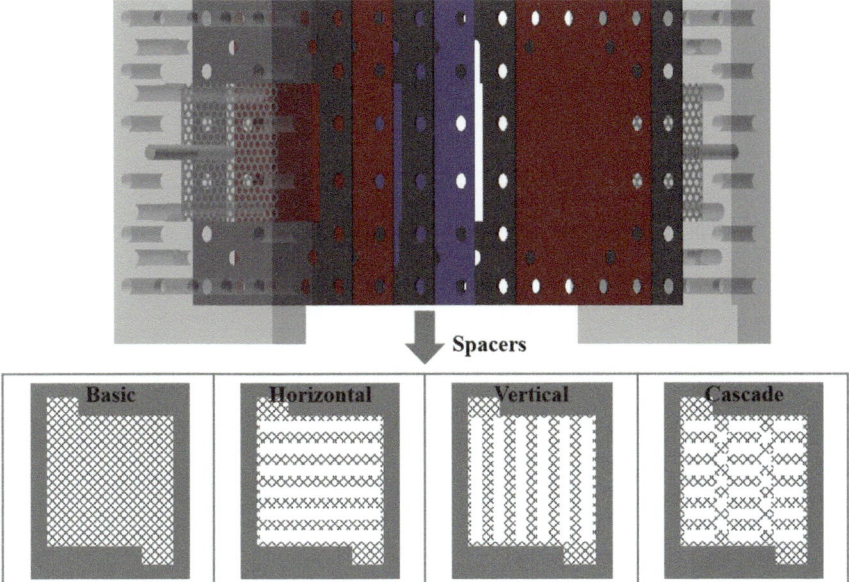

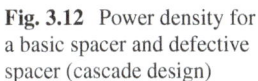

Fig. 3.11 Possible designs for defective spacers

Fig. 3.12 Power density for a basic spacer and defective spacer (cascade design)

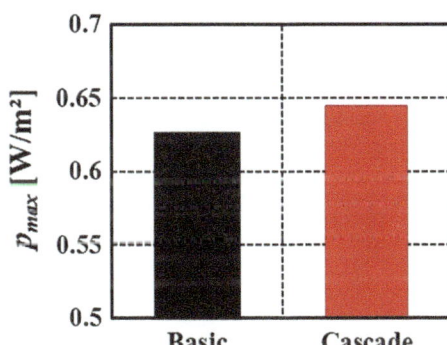

We compared a basic spacer without any defect with a defective spacer (cascade design) and found a slight increase in the power density for the defective spacer (Fig. 3.12).

Reference

D.A. Vermaas, M. Saakes, K. Nijimeijer, Power generation profiled membranes in reverse electrodialysis. J. Membr. Sci. **385–386**, 234–242 (2011)

Chapter 4
RED Applied to Desalination

4.1 Brine Recovery Using RED in Membrane-Based Desalination

RED performance can be improved by using highly concentrated brines in the desalination process. Desalination brines have high chemical potentials because of ions that remain after water is eliminated from seawater. To date, there are only a few studies on using brines as the concentrated solution in RED systems. This chapter shows the power enhancement of RED using brines discharged from two different membrane-based desalination processes: (1) reverse osmosis (RO), the most popular method in the desalination market, and (2) forward osmosis (FO), a novel concept without energy-intensive high-pressure processes. The effects of desalination brines on RED performance were evaluated both experimentally and numerically. We modified the one dimensional RED model developed by Veerman et al. (2011) and successfully validated this modified numerical model with lab-scale RED stacks. We characterized the maximum power density and net power density of RED according to the intermembrane distance and inlet flow rate. We assumed two different realistic conditions for the diluted solutions using river water and seawater. We also analyzed the combined effects of integrating RED with the desalination processes. We found a reduction in specific energy consumption (SEC) owing to the chemical energy recovery using RED.

4.1.1 Process Description

Figure 4.1 shows a schematic diagram of the combined desalination and RED processes. We considered two different membrane-based desalination processes. RO produces freshwater using high pressure above osmotic pressure, whereas FO natu-

© The Author(s), under exclusive license to Springer Nature Singapore Pte Ltd. 2019   23
D. Kim et al., *Energy Generation using Reverse Electrodialysis*,
SpringerBriefs in Applied Sciences and Technology,
https://doi.org/10.1007/978-981-13-0314-2_4

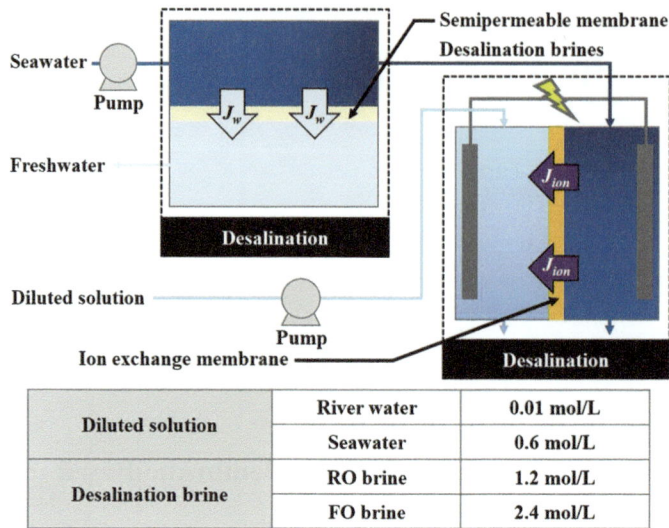

Diluted solution	River water	0.01 mol/L
	Seawater	0.6 mol/L
Desalination brine	RO brine	1.2 mol/L
	FO brine	2.4 mol/L

Fig. 4.1 Combined membrane-based desalination and RED processes

rally obtains freshwater by employing so-called draw solutions, which have higher osmotic pressures than seawater. As we considered two locations (estuary and coastal area) for these desalination plants, we simulated both river water and seawater as the diluted solution for the RED processes.

We assumed the concentrations of seawater (~35,000 mg/L TDS) and river water (100–600 mg/L TDS) to be 0.6 and 0.01 mol/L NaCl, respectively. In the RO process, the water recovery rate, which is defined as the outlet flow rate per inlet flow rate, is typically reported at ~50%; therefore, we fixed the concentration of the RO brine at twofold that of seawater (1.2 mol/L). Previous studies have reported that the water recovery rate is expected to reach 75% in FO, and we consequently set the concentration of the FO brine to 2.4 mol/L.

4.1.2   RED Modeling

Most RED systems are fabricated using a plate and frame type arrangement, where a unit cell is serially stacked (Vermaas et al. 2011a, b; Güler et al. 2013). The maximum voltage (i.e., the OCV) is a linear function of the number of cell pairs, and the maximum current, called the short circuit current (SCC), is proportional to the area of the working membrane. For this study, we modified the one-dimensional model proposed by Veerman et al. (2011) to analyze the effect of desalination brines. This model is based on a unit cell that consists of an AEM, a concentrated compartment, a CEM, and a diluted compartment. Figure 4.2 shows an illustration of a RED model

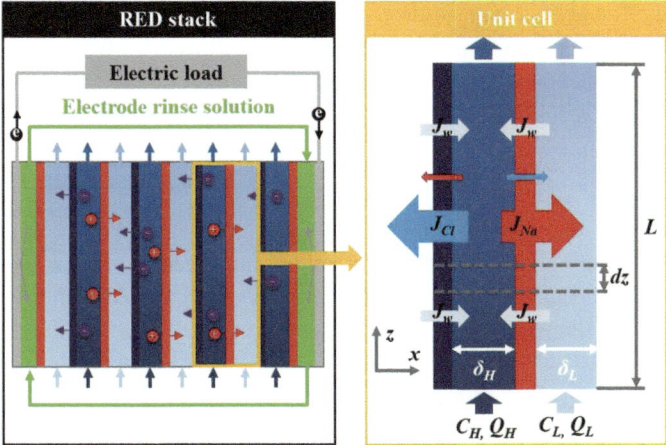

Fig. 4.2 Illustration of RED stacks with a plate and frame arrangement and a unit cell model with length L

based on a unit cell with length L and width W. The concentrated and diluted solutions enter each compartment with flow rate Q and concentration C. There are several assumptions in this model: (1) the effect of ions other than sodium and chloride ions is neglected in all solutions; (2) a concentration change is considered solely along the flow direction; (3) the transference number and resistance of the IEMs are fixed as constant values, as suggested by Güler et al. (2013); (4) the inlet flow rate is uniform in each compartment; (5) the influence of the electrode rinse solution at both ends of the stack is not considered; and (6) the overall RED power generation is linearly proportional to the number of cells.

In RED, an electric circuit is created by ion transport from the concentrated compartment to the diluted compartment across the IEMs. The concentration difference significantly affects the voltage, current, and power of this circuit. When the external circuit between the anode and the cathode is separated, the RED voltage is maximized because of charge accumulation on the electrodes. The maximum voltage can be expressed by the Nernst equation with the transference number (α) determining the permselectivity of the IEMs (Tedesco et al. 2012; Veerman et al. 2011):

$$V_{oc}(z) = (\alpha_{CEM} + \alpha_{AEM}) \frac{RT}{zF} \ln \left(\frac{\gamma_H(z)C_H(z)}{\gamma_L(z)C_L(z)} \right) \tag{4.1}$$

where R is the universal gas constant (8.314 J mol^{-1} K^{-1}), T is the ambient temperature, z is the valence number, F is the Faraday constant (96,485 C mol^{-1}), γ is the molar activity coefficient, and C is the concentration. Subscripts H and L refer to the concentrated and diluted solutions, respectively.

A current can be generated by connecting an external circuit to the RED system. In the ideal case without internal resistance, there would be no limit on the current generated. However, a voltage drop occurs with current generation because there are various types of resistance in RED. Typically, there are two types of internal resistance: ohmic and nonohmic resistance (Vermaas et al. 2012). Ohmic resistance includes the membrane and solution resistance and is expressed as:

$$R_{ohmic}(z) = \frac{1}{\beta}\left(R_{CEM} + R_{AEM} + \frac{\delta_H}{\Lambda_H(z)C_H(Z)} + \frac{\delta_L}{\Lambda_L(z)C_L(Z)}\right) \quad (4.2)$$

where R is the resistance; β is the open ratio of the spacer; δ_H and δ_L are the thicknesses of the concentrated and diluted compartments, respectively; and Λ is the molar conductivity of the working solutions.

Nonohmic resistance is mainly determined by two factors in RED [37]. One factor is the change in concentration of both solutions along the flow direction. This effect can be directly computed in our model. The other is the concentration polarization (CP) at the membrane surface (Vermaas et al. 2012), which is usually influenced by channel length, flow rate, and flow structure. The presence of spacers, which create tortuous flow, can reduce the influence of CP by providing extra flow mixing within each compartment (Vermaas et al. 2014). We used the following empirical equation proposed (Vermaas et al. 2012) to consider the effect of CP in a RED system with spacers:

$$R_{CP} = 0.31 t_H \frac{\delta_H}{L} + 0.31 t_L \frac{\delta_L}{L} + 0.05, \quad (4.3)$$

where t_H and t_L are the residence times in the concentrated and diluted compartments and L is the channel length.

The current can be defined as

$$I(z) = \frac{V_{OC}(z)}{R_{int}(z) + R_{load}}, \quad (4.4)$$

where R_{int} is the internal resistance, including ohmic and CP resistance, and R_{load} is the external resistance.

The current density is computed by integrating Eq. (4.4) with respect to z and dividing by the overall working area:

$$i = \frac{\sum_{z=0}^{L} I(z)dz}{2LW}. \quad (4.5)$$

The power output P and power density p (power per active membrane area) can be evaluated as

$$P(z) = I(z)^2 R_{load} \quad (4.6)$$

$$p = \frac{\sum_{z=0}^{L} P(z)dz}{2LW}. \quad (4.7)$$

It is important to accurately determine thermodynamic properties for the working solutions to reduce computation errors. We calculated the relevant properties (e.g., the molar activity coefficient, viscosity, dielectric constant, and molar conductivity) of sodium chloride solutions using accurate correlations. The molar activity coefficient associated with the OCV has a value between 0 and 1, with solutions displaying ideal behavior when this value is close to 1. There are several theoretical approaches to estimate the molar activity coefficient, including the Debye–Hückel equation, the extended Debye–Hückel equation, and the Pitzel equation. Among these, we chose the Pitzel equation because it is known to be in excellent agreement with experimental data for the concentration range of 0–6 mol m^{-3} at a temperature of 25 °C:

$$\ln \gamma_{Nacl}(z) = -A_1 \left[\frac{\sqrt{I}}{1 + 1.2\sqrt{I}} + \frac{2}{1.2} \ln(1 + 1.2\sqrt{I})1 + 1.2\sqrt{I} \right] + m B^{\gamma} + m C^{\gamma},$$

(4.8)

where A_1 is the modified Debye–Hückel constant (0.3915 at 25 °C), I is the ionic strength, and m is the molarity [mol m^{-3}] of the solution. The terms B^{γ} and C^{γ} are the Pitzel parameters; details of these parameters can be found in Tedesco et al. (2012).

Molar conductivity influences the internal resistance in RED. We used the equation suggested by Islam et al. [44], which is an extension of the Falkenhagen–Leist–Kelbg (FLK) equation, to calculate the molar conductivity of the electrolyte at high concentration:

$$\Lambda(z) = \left(\Lambda_0 - \frac{B_2'(C)\sqrt{C}}{1 + B'(C)a\sqrt{C}} \right) \left(1 - \frac{B_1'(C)\sqrt{C}}{1 + B'(C)a\sqrt{C}} F'(C) \right),$$

(4.9)

where Λ_o is the equivalent conductivity of the sodium chloride solution at infinite dilution and a is the correlation factor to fit the experimental data (3.79 Å). The $B'(C)$, $B_1'(C)$, $B_2'(C)$, and $F'(C)$ are as follows:

$$B'(C) = \frac{50.29 \times 10^8}{\sqrt{\varepsilon(z)T}},$$

(4.10)

$$B_1'(C) = \frac{2.5}{\mu(z)\sqrt{\varepsilon(z)T}},$$

(4.11)

$$B_2'(C) = \frac{8.204 \times 10^6}{\sqrt[3]{\varepsilon(z)T}}, \quad \text{and}$$

(4.12)

$$F'(C) = \frac{\left[\exp\left(0.2929 B'a\sqrt{C(z)} - 1\right)\right]}{0.2929 B'a\sqrt{C(z)}},$$

(4.13)

where μ and ε are viscosity and permittivity, respectively.

In most previous studies on RED modeling, the variation in the viscosity and permittivity were not considered or were assumed to be a linear function of the concentration. We employed an empirical model for these parameters to improve accuracy. We applied the extended Jones–Dole equation and the equation suggested by Buchner et al. (1998) for a more precise computation of the viscosity and dielectric constant of our sodium chloride solutions:

$$\mu(z) = \mu_0 \left(1 + 0.0061\sqrt{C} + 0.078C + 0.013C^2\right) \text{ and} \qquad (4.14)$$

$$\varepsilon(z) = \varepsilon_0 - 15.2C + 3.64C^2, \qquad (4.15)$$

where μ_0 and ε_0 are the viscosity and permittivity of water.

The solution concentrations vary along the longitudinal direction (z-direction) owing to the lateral transport (y-direction) of solutes and solvents. These longitudinal migrations are closely related to the previously mentioned electric variables in RED. The solute transport (J_S) from the concentrated compartment to the diluted compartment is divided into counter-ion (opposite charge as at the IEM surface and related to the current density) and co-ion transport, represented by

$$J_s = \frac{i(z)}{F} + \left(\frac{D_{NaCl}}{h_{CEM}} + \frac{D_{NaCl}}{h_{AEM}}\right)(C_H(z) - C_L(z)), \qquad (4.16)$$

where D_{NaCl} is the permeability coefficient of the solute, which is set to a constant value of 10^{-12} m^2 s^{-1} (Tedesco et al. 2015); h_{CEM} is the thickness of the CEM; and h_{AEM} is the thickness of the AEM.

The water transport resulting from osmosis is in the direction opposite to the ion transport. Equation (4.17) presents the molar flux of water:

$$J_w(z) = -\left(\frac{D_w}{h_{CEM}} + \frac{D_w}{h_{AEM}}\right)(C_H(z) - C_L(z)), \qquad (4.17)$$

where D_w represents the diffusion coefficient of water through the membranes ($D_w = 10^{-9}$ m^2 s^{-1}).

The overall mass balances for sodium chloride in an infinitesimal control volume can be expressed as

$$\frac{dC_H}{dz} = -\frac{W}{Q_H} J_s(z) + C_H(z)\frac{W}{Q_H} J_w(z) V_w \text{ and} \qquad (4.18)$$

$$\frac{dC_L}{dz} = \frac{W}{Q_L} J_s(z) - C_L(z)\frac{W}{Q_L} J_w(z) V_w, \qquad (4.19)$$

where V_w is the molar density of water.

We solved this system of differential equations (Eqs. 4.18 and 4.19) with the inlet boundary conditions using the backward finite difference approximation with an integration step of dz $=$ L/3000. We first assigned initial values for the concentration, external resistance, intermembrane distance, and inlet flow rate, and then obtained the local values of the electrical variables, solute flux, and water flux in each step. The solution of all the aforementioned equations gives the variation in the concentration of the working solutions along the flow direction for the next step. The voltage was averaged over the whole step. We finally computed the current and power generated in RED for the full range.

4.1.3 Model Validation

We first compared the current–voltage (I–V) characteristics, known as the polarization curve, of RED between the analytical and experimental results to verify our RED model. The experimental measurements were performed in the so-called galvanostatic mode, where the terminal voltage is measured from the change in the current. We increased the current by steps of 20 mA each minute. The polarization curves were measured over at least 10 min. The operating temperature and the number of cell pairs were fixed at ~25 °C and 5 °C, respectively, in all RED experiments. We measured each data point at least four times to ensure the repeatability of the data. The student t-distribution with a 95% confidence interval was used to determine error bars, based on these independent measurements. Figure 4.3a shows the polarization curves for our analytical and experimental results. The NaCl concentration of the concentrated and diluted solutions was 0.01 mol/L (modeled river water) and 0.6 mol/L (modeled seawater). We fixed the flow rate of both streams at 10 mL/min per compartment and the intermembrane distance of each compartment was 0.2 mm. We found almost perfect agreement between the open circuit voltages from the modeling (~0.908 V) and the experiment (~0.906 V). Both polarization curves show ohmic behavior where the voltage linearly decreases with increasing current density. This behavior is significantly different from that of a fuel cell, which shows three clear regimes of loss: activation loss, ohmic loss, and concentration losses. The overall polarization curve shows good accordance between the experimental and model data over the whole range of current densities.

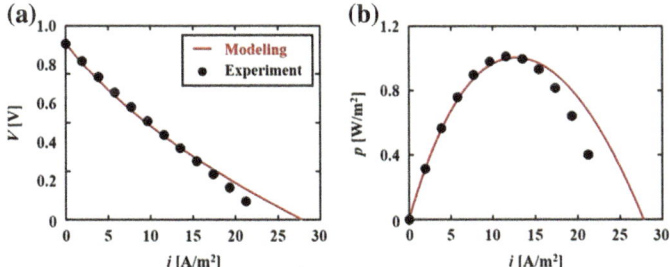

Fig. 4.3 Comparison of **a** polarization and **b** power density curves for the analytical model (—) and experimental data (●): C_H and C_L = 0.6 and 0.01 mol/L, δ = 0.2 mm, Q_{cell} = 10 mL/min, and N_{cell} = 5

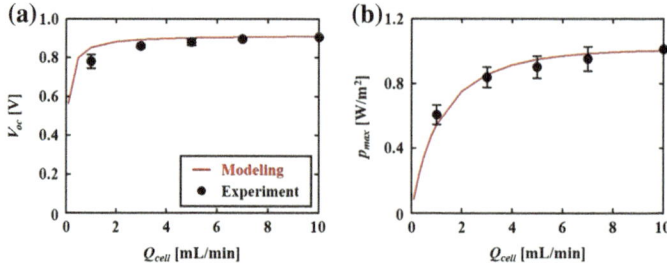

Fig. 4.4 **a** OCV and **b** maximum power density of experimental (●) and simulated data (—) at inlet flow rates ranging from 0.1 to 10 mL/min: C_H and C_L = 0.6 mol/L and 0.01 mol/L, δ = 0.2 mm, and N_{cell} = 5

Figure 4.3b shows the power density curves. We obtained the power density of RED from the polarization curve using the formula $p = V \cdot i$. The power density curve has a parabolic shape because the I–V characteristics showed a linear trend. There was a little difference between the maximum power density from the modeling (1.00 W/m^2) and that from the experiment (1.01 W/m^2).

The concentration in each compartment varied with the inlet flow rate because of the variation in the residence time. This caused a change in the concentration difference along the flow direction. For that reason, the OCV, which depends on the concentration difference between the concentrated and diluted solutions, can be influenced by the inlet flow rate. We analyzed the effect of the inlet flow rate on the OCV using the RED model. Figure 4.4a shows the OCV for inlet flow rates between 0.1 and 10 mL/min. The inlet flow rates of both compartments were varied identically in this model. Other conditions were fixed with a diluted solution of 0.01 mol/L, a concentrated solution of 0.6 mol/L, and an intermembrane distance of 0.2 mm. Under these conditions, the OCV dramatically increased until a flow rate of 2 mL/min and then gradually converged toward the theoretical maximum. We compared this analytically obtained profile with the experimental results for five data points obtained with flow rates of 1, 3, 5, 7, and 10 mL/min. The measured OCV values are within 8% of the calculated values for each data point. We also compared the power density of the numerical and experimental results using the same conditions. These results are in good agreement, as shown in Fig. 4.4b.

4.1.4   Effect of Desalination Brines

We numerically and experimentally evaluated the influence of two desalination brines on RED power generation with two different diluted solutions. We assumed the concentration of the RO and FO brines to be two and four times higher than that of seawater, respectively, based on the water recovery rate of each process. The intermembrane distance and the inlet flow rate were 0.2 mm and 10 mL/min, respec-

Fig. 4.5 Measured (■) and calculated (■) maximum power density for various concentrated solutions (seawater, 0.6 mol/L; RO brine, 1.2 mol/L; FO brine, 2.4 mol/L) with **a** river water (0.01 mol/L) and **b** seawater as the diluted solution: $\delta =$ 0.2 mm and $Q_{cell} =$ 10 mL/min

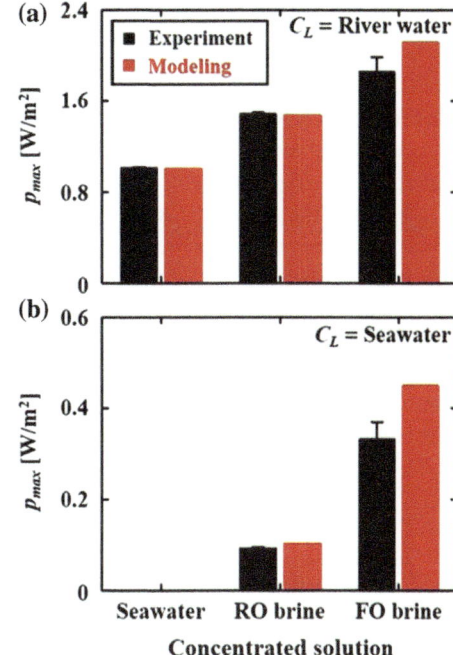

tively. Figure 4.5a shows a comparison of the maximum power density by varying the concentrated solutions against a fixed diluted solution (0.01 mol/L). In the thermodynamic model based on Gibbs energy, the maximum energy generation obtained with perfect mixing is linearly proportional to the concentration of the concentrated solution for a fixed diluted solution. The trend in the numerical results nearly agrees with that of the thermodynamic analysis. We found that the growth rate of the power density was half that of the concentration of the concentrated solution because of the decrease in the mixing rate as the concentration increases. The experimental results show good agreement with the numerical results, thereby agreeing with the thermodynamic analysis. The effect of the concentration variation was well predicted by the RED model, although there was a slight discrepancy (within 10%) for the FO brine. The maximum power densities with the modeled RO brine and the modeled FO brine were 1.48 and 1.86 W/m^2, respectively. We also analyzed the power density when seawater was used as the diluted solution. The power densities were significantly lower than when river water was used owing to the decrease in the concentration difference, as shown in Fig. 4.5b. We experimentally obtained power densities of 0.09 W/m^2 (modeled RO brine) and 0.37 W/m^2 (modeled FO brine).

4.2   Modeling RO Integrated with RED: Evaluation of Predesalination and Brine Recovery

In this chapter, we consider the integration of the RED device to the RO desalination process. RED can be used for predesalination before the RO module and the chemical energy recovery using brine discharged from the RO module (Fig. 4.6).

We used the one-dimensional mathematical model proposed by Geraldes et al. (2005) to analyze the RO module. We first conducted the analysis on the variation of the flow rate, the pressure, the water permeability, and the concentration in the feed channel along with the flow direction (z-direction). Figure 4.7 shows the results under fixed inlet conditions.

Figure 4.8 shows the SEC with the feed flow rate and feed pressure. The SEC monotonously increases with the feed flow rate at the fixed feed pressure. On the other hand, it shows a non-monotonic trend with the feed pressure at a flow rate of 1.75 L/s. The minimal SEC was obtained at a feed pressure of 5.8 MPa.

4.2.1   Integration with RED

We performed an evaluation of the RO desalination process integrated with RED. Figure 4.9 shows the results for the predesalination, chemical energy recovery, and predesalination/the chemical energy recovery. The SEC to produce freshwater was

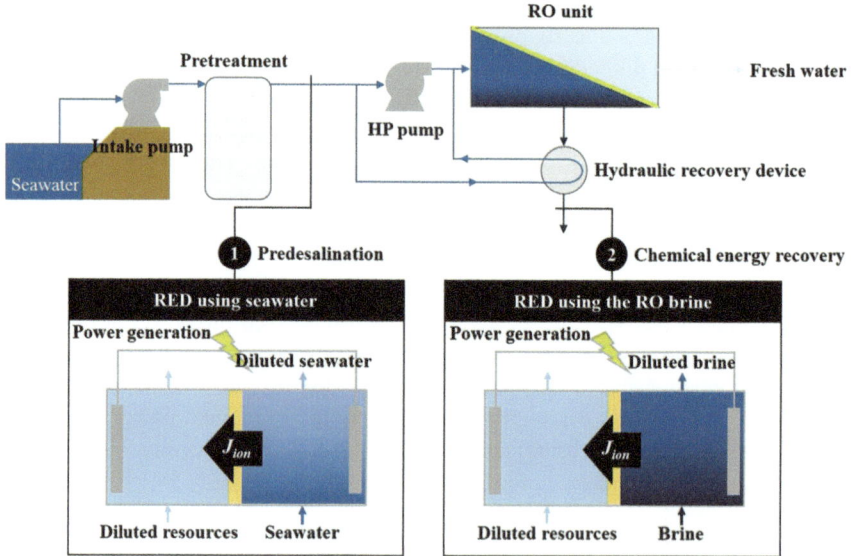

Fig. 4.6 Schematics of the RO desalination process integrated with the RED device

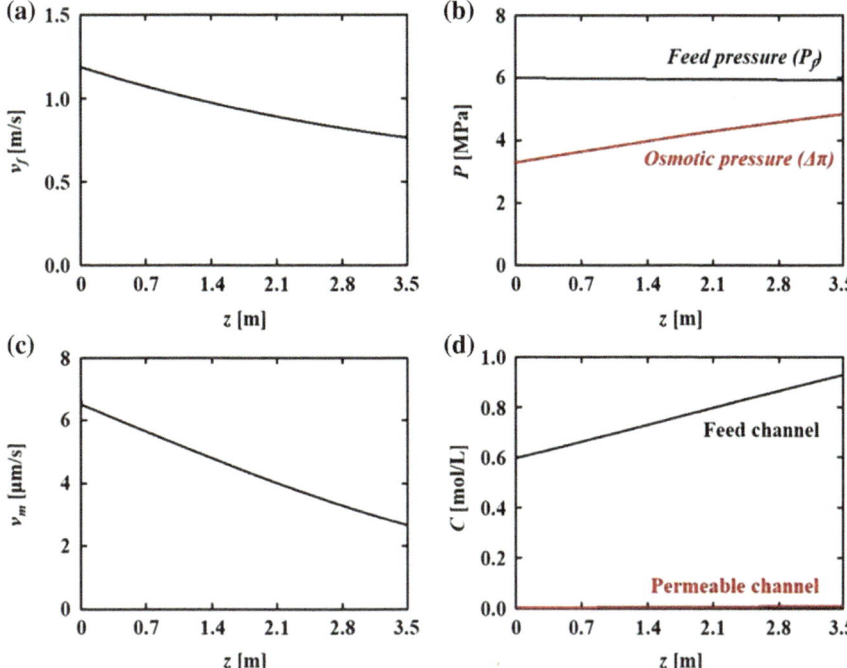

Fig. 4.7 Change in **a** velocity, **b** pressure, **c** water permeability, and **d** concentration along the longitudinal direction (z-direction): $N_{module} = 4$, $Q_f = 1.75$ L/s (vf = ~1.2 m/s), PHP = 6 MPa, and CS = ~0.6 mol/L

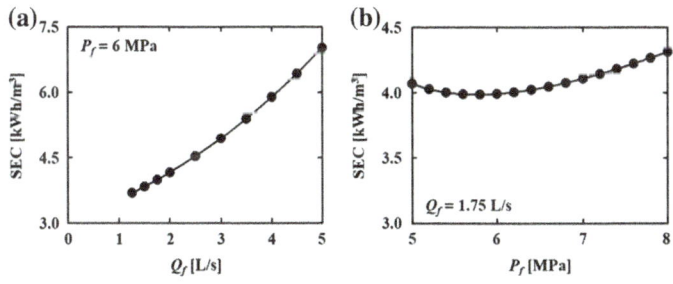

Fig. 4.8 SEC as a function of **a** flow rate ($P_f = 6$ MPa) and **b** feed pressure ($Q_f = 1.75$ L/s)

significantly reduced when the RED was used as the predesalination module. The reduction rate was 25.4% at maximum.

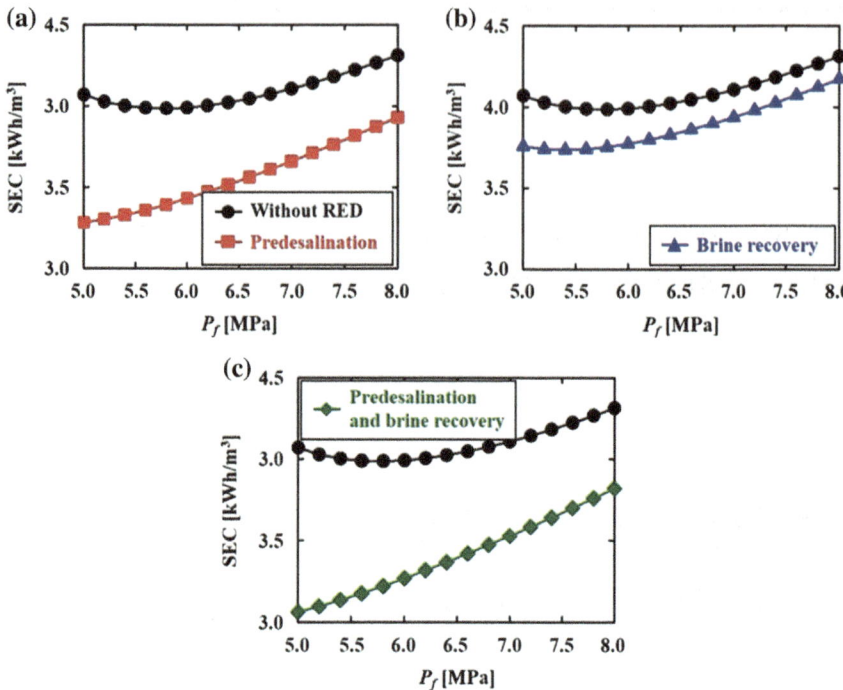

Fig. 4.9 Reduction in the SEC using RED: **a** predesalination, **b** chemical energy recovery, and **c** predesalination and chemical energy recovery

References

R. Buchner, G.T. Hefter, P.M. May, Dielectric relaxation of aqueous NaCl solutions. J. Phys. Chem. A **103**, 1–9 (1998)

V. Geraldes, N.E. Pereira, M.N. de Pinho, Simulation and optimization of medium-sized seawater reverse osmosis processes with spiral-wound modules. J. Ind. Eng. Chem. **44**, 1897–1905 (2005)

E. Güler, R. Elizen, D.A. Vermaas, M. Saakes, K. Nijmeijer, Performance-determining membrane properties in reverse electrodialysis. J. Membr. Sci. **446**, 266–276 (2013)

M. Tedesco, A. Cipollina, A. Tamburini, W. van Baak, G. Micale, Modelling the reverse electrodialysis process with seawater and concentrated brines. Desalination Water Treat. **49**, 404–424 (2012)

M. Tedesco, A. Cipollina, A. Tamburini, D.L. Bogle, G. Micale, A simulation tool for analysis and design of reverse electrodialysis using concentrated brines. Chem. Eng. Res. Des. **93**, 441–456 (2015)

J. Veerman, M. Saakes, S.J. Metz, G.J. Harmsen, Reverse electrodialysis: a validated process model for design and optimization. Chem. Eng. J. **166**, 256–268 (2011)

D.A. Vermaas, M. Saakes, K. Nijimeijer, Doubled power density from salinity gradient at reduced intermembrane distance. Environ. Sci. Technol. **45**, 7089–7095 (2011a)

D.A. Vermaas, M. Saakes, K. Nijimeijer, Power generation profiled membranes in reverse electrodialysis. J. Membr. Sci. **385–386**, 234–242 (2011b)

D.A. Vermaas, E. Güler, M. Saakes, K. Nijmeijer, Theoretical power density from salinity gradients using reverse electrodialysis. Energy Procedia **20**, 170–184 (2012)

D. Vermaas, M. Saakes, K. Nijmeijer, Enhanced mixing in the diffusive boundary layer for energy generation in reverse electrodialysis. J. Membr. Sci. **453**, 312–319 (2014)

Chapter 5
Parametric Study of RED Using Ammonium Bicarbonate Solution to Recover Low-Grade Waste Heat

Improving the efficiency of energy conversion systems (e.g., thermal power plants and vehicles) is increasingly important owing to growing concern regarding oil depletion and the environmental pollution resulting from byproducts of burning of fossil fuels. Currently, the rate of transforming useful output from primary energy resources is approximately 40% [95]. The rest is lost in operation or released as thermal energy. Accordingly, technologies to recover waste heat discharged to environmental surroundings have received a significant interest over the past decades. Several cases, including combined heat and power plants (CHP), have been successfully applied in industrial fields. Recently, a new method, the thermal-driven electrochemical generator (TDEG), is proposed. Figure 5.1 shows the schematic of TDEG. Here, we present an evaluation on the RED-NH_4HCO_3 (Ammonium bicarbonate) system to produce power using the waste heat recovery. We investigated various parameters including the concentration difference, IEM, inlet flow rate, and intermembrane distance. We also computed the net power with consideration of the pressure drop resulting in the external loss.

We first compared RED performance between a sodium chloride solution and an ammonium bicarbonate solution. Figure 5.2 shows the results of the i-V characteristics, including the polarization curve and power density curve for a temperature of ~ 25°C. The black line (●) represents RED performance with the sodium chloride solution and the red line (■) denotes RED performance with the ammonium bicarbonate solution. The concentration of the concentrated and the diluted solutions were fixed at 0.6 mol/L and 0.01 mol/L, respectively. Other conditions were an intermembrane distance of 0.3 mm and an inlet flow rate of 10 mL/min per unit cell. The polarization curve was measured in the galvanostatic mode, where the current is swept. The current increased in 10 mA steps per minute. We conducted the experiment at least three times for each data point to ensure the repeatability of the data. The polarization curve is the most ubiquitous characterization techniques for electrochemical cells such as batteries and fuel cells. It typically has three clear regimes in fuel cells: activation loss at low current, ohmic loss owing to internal resistance, and

© The Author(s), under exclusive license to Springer Nature Singapore Pte Ltd. 2019   37
D. Kim et al., *Energy Generation using Reverse Electrodialysis*,
SpringerBriefs in Applied Sciences and Technology,
https://doi.org/10.1007/978-981-13-0314-2_5

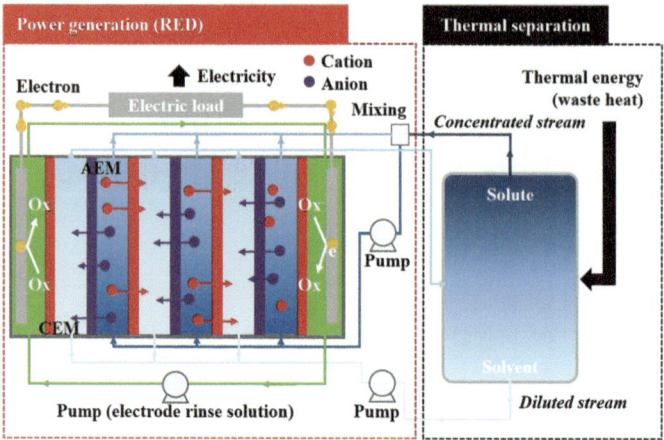

Fig. 5.1 Schematic illustration of RED integrated with a regeneration device

concentration loss at high current. In contrast, RED shows ohmic behavior where the voltage linearly drops with the current, as shown in Fig. 5.2. This is because the electrode rinse solution circulated between both ends plays a role in reducing the loss of redox reactions on the electrodes. We calculated the power density of RED from the polarization curve using the simple formula $p = V \cdot i$, where p, V, and i represent the power density, terminal voltage, and current density, respectively. The power density curve shows a parabolic shape; consequently, maximum power is obtained at the center of the current density. The maximum power of the ammonium bicarbonate solutions (0.42 W/m^2) was lower than that of the sodium chloride solution (0.56 W/m^2).

The variation in the concentration difference can considerably affect the electrochemical performance of RED. We evaluated RED performance with the change in the concentration of the concentrated and the diluted solutions. Figure 5.3 shows the results on the OCV and the maximum power densities as a function of the concentration of the concentrated solutions. We tested four concentrated solutions of 0.6, 0.9, 1.2, and 1.5 mol/L, although the solubility of ammonium bicarbonate can be up approximately 2 mol/L at room temperature. The concentration of the diluted solution was set at 0.01 mol/L. The flow rate and the intermembrane distance were 10 mL/min and 0.3 mm, respectively. The OCV and power density gradually increased as the concentration of the concentrated solution increased.

Figure 5.4 shows the OCV and power density as a function of the concentration of the diluted solution. Four diluted solutions of 0.005, 0.01, 0.015, and 0.02 mol/L were determined for the evaluation of RED performance at a fixed concentrated solution of 1.5 mol/L. The flow rate and the intermembrane distance were 10 mL/min and 0.03 mm, respectively. The OCV trend showed a monotonous decrease with increasing concentration of the diluted solution from 0.005 mol/L to 0.02 mol/L because of the reduction in the concentration difference. Internal resistance, including

Fig. 5.2 a Polarization and
b power density curve for a
sodium chloride solution (●)
and an ammonium
bicarbonate solution (■): C_H
= 0.6 mol/L, C_L =
0.01 mol/L, Q_{cell} =
10 mL/min, and δ = 0.3 mm

Fig. 5.3 a OCV and
b maximum power density
as a function of the
concentration of
concentrated solution at a
fixed concentration of
diluted solution: C_L =
0.01 mol/L, Q_{cell} =
10 mL/min, and δ = 0.3 mm

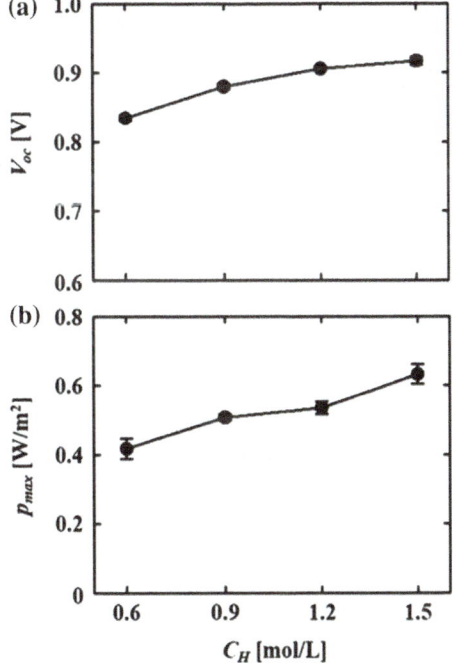

Fig. 5.4 **a** OCV and
b maximum power density
as a function of the
concentration of diluted
solution at a fixed
concentration of
concentrated solution: $C_H =$
1.5 mol/L, $Q_{cell} =$
10 mL/min, and $\delta = 0.3$ mm

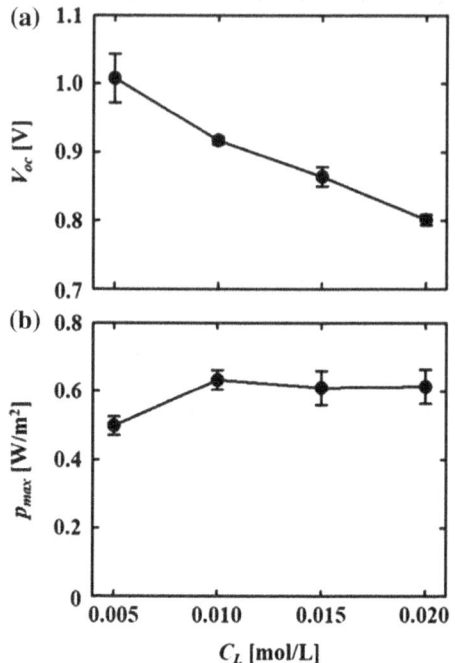

Fig. 5.4 **a** OCV and **b** maximum power density as a function of the concentration of diluted solution at a fixed concentration of concentrated solution: $C_H =$ 1.5 mol/L, $Q_{cell} =$ 10 mL/min, and $\delta = 0.3$ mm

membrane and compartment resistances, as well as the OCV affects power generation in the RED system. Typically, the power generation depends on the concentration of the diluted compartment owing to the low solution conductivity. Accordingly, the lower concentration of the diluted solution does not ensure high power in the RED system. In our experiments, power density had an optimal value at 0.01 mol/L. Our maximum point was slightly different from that in Luo et al. 2012 0.02 mol/L). We believe that this is due to the different system configurations; for example, the choice of IEMs and geometries.

We compared RED performance using two different commercial IEMs under identical conditions: the heterogeneous membrane (Excellion I-100 and I-200) and the homogenous membrane (Neosepta CMX and AMX). The nonconducting inert fraction usually reported that the heterogeneous membrane is significantly greater than that of the homogenous membrane. Consequently, we found that the power density with the Neosepta membranes (0.63 W/m^2) was 2.3 times higher than that with the Excellion membranes (0.27 W/m^2) for an inlet flow rate of 10 mL/min and an intermembrane distance of 0.3 mm, as presented in Fig. 5.5.

Finally, we evaluated the effect of intermembrane distance on RED power. We employed four different spacers of 0.2, 0.3, 0.5, and 0.8 mm. Other conditions were an inlet flow rate of 10 mL/min, a concentrated solution of 1.5 mol/L, and a diluted solution of 0.01 mol/L. The black line (■), red line (●), and dotted line denote the maximum power density, net power density, and pumping loss, respectively in Fig. 5.6. Vermaas et al. (2011) reported that power density is substantially improved

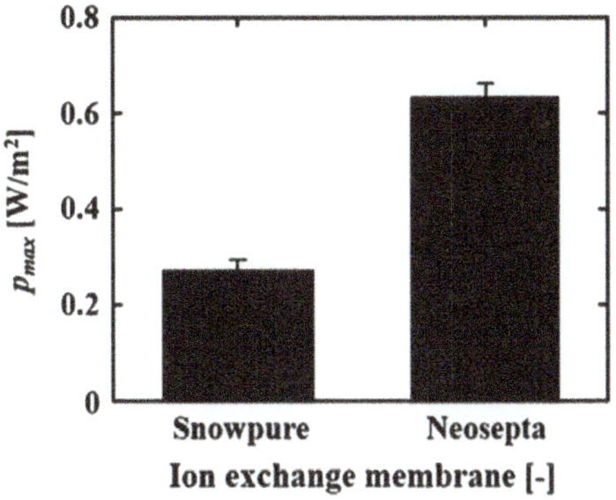

Fig. 5.5 Power densities for two different IEMs: $C_H =$ 1.5 mol/L, $C_L = 0.01$ mol/L, $Q_{cell} = 10$ mL/min, and $\delta =$ 0.3 mm

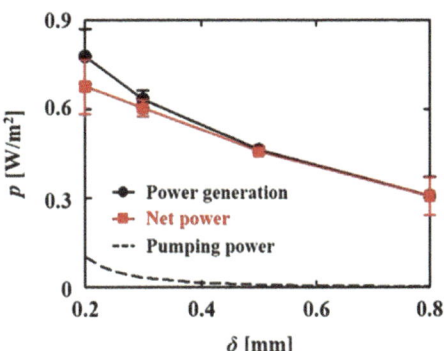

Fig. 5.6 Power density as a function of intermembrane distance (δ): $C_H =$ 1.5 mol/L, $C_L = 0.01$ mol/L, and $Q_{cell} = 10$ mL/min

with a decrease in the intermembrane distance because of the reduced internal resistance of compartments. The power density monotonously increased when the intermembrane distance decreased. We obtained the highest power density (0.77 W/m²) at an intermembrane distance of 0.2 mm. The net power density (0.62 W/m²) was also maximized at the intermembrane distance of 0.2 mm.

Reference

X. Luo, X. Cao, Y. Mo, K. Xiao, X. Zhang, P. Liang, X. Huang, Power generation by coupling reverse electrodialysis and ammonium bicarbonate: implication for recovery of waste heat. Electrochem. Commun. **19**: 25–28 (2012)

D.A. Vermaas, M. Saakes, K. Nijimeijer, Doubled power density from salinity gradient at reduced intermembrane distance. Environ. Sci. Technol. **45**, 7089–7095 (2011)

Chapter 6
Nanofluidic RED

The appearance of novel technologies such as microelectromechanical systems (MEMs) in the manufacturing sector has facilitated the fabrication of nanometer-sized structures, and new research fields have emerged in this regard. Nanofluidics, which is the study of fluid flows in a channel or pore with at least one characteristic dimension below 100 nm, is one such field of research. Several unique features caused by the extremely high surface-to-volume ratio of nanofluidics have been identified (Zangle et al. 2010; Kim et al. 2010a). When a channel is in contact with an aqueous solution, the channel surface is typically charged owing to surface ionization, ion adsorption, or ion dissolution. Counter-ions are dragged onto the channel surface and co-ions are repelled. An electric double layer (EDL) is thus formed on the channel surface and can be divided into the Stern and diffuse layers. A bulk layer exists in a micrometer-sized channel but is eliminated in a nanometer-sized channel owing to strong electrostatic interaction between the charged surface and ions in the solution. Consequently, the nanochannel may have permselectivity (also known as ion selectivity). These characteristics have been utilized in various applications, such as sample preconcentration/separation (Ko et al. 2012), desalination (Kim et al. 2010b), mixing (Kim et al. 2008; Lee and Kim 2012), and energy harvesting (Eijkel and van den Berg 2010; Kim et al. 2010a).

Figure 6.1a depicts the mechanism and an equivalent circuit for an NRED system. When more and less concentrated solutions are separately supplied to each end of the nanochannel, counter-ions and co-ions diffuse from the concentrated solution to the diluted solution. The counter-ions migrate more easily than do the co-ions because of an electrostatic interaction inside the nanochannel. This asymmetric ion transport causes electrochemical redox reactions on electrode surfaces to maintain electroneutrality in the solution. Accordingly, electrons can be transferred to an external electric circuit, and electric power is produced with an electric load. The equivalent circuit of the NRED system is presented in Fig. 6.1b. E_{redox} is the potential on the electrodes resulting from the unequal voltage drops in different electrolyte concentrations, and EEMF is the potential generated from the concentration gradient

© The Author(s), under exclusive license to Springer Nature Singapore Pte Ltd. 2019   43
D. Kim et al., *Energy Generation using Reverse Electrodialysis*,
SpringerBriefs in Applied Sciences and Technology,
https://doi.org/10.1007/978-981-13-0314-2_6

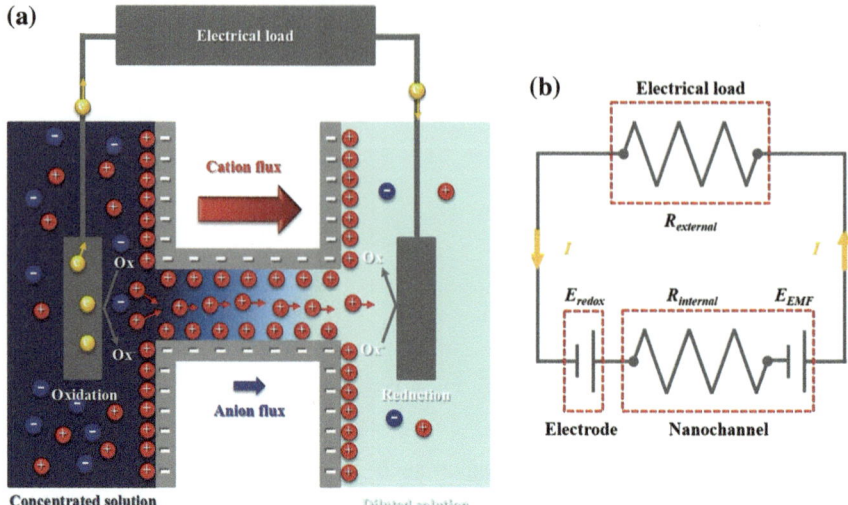

Fig. 6.1 **a** Working principle of the NRED system and **b** equivalent circuit, which includes the electromotive force by asymmetric ion transport and internal resistance on the solution and nanochannel, redox potential at the electrodes, and load resistance

between both ends of the nanochannel. $R_{external}$ and $R_{internal}$ are electrical resistances of the load and NRED system, respectively. $R_{internal}$ includes electrical resistance caused by the nanochannel and solutions.

References

J.C.T. Eijkel, A. van der Berg, Nanofluidics and the chemical potential applied to solvent and solute transport. Chem. Soc. Rev. **39**, 957–973 (2010)

D. Kim, A. Raj, L. Zhu, R.I. Masel, M.A. Shannon, Non-equilibrium electrokinetic micro/nano fluidic mixer. Lab Chip **8**, 625–628 (2008)

D.K. Kim, C. Duan, Y.F. Chen, A. Majumdar, Power generation from concentration gradient by reverse electrodialysis in ion-selective nanochannels. Microfluid. Nanofluid. **9**, 1215–1224 (2010a)

J. Kim, S.H. Ko, K.H. Kang, J. Han, Direct seawater desalination by ion concentration polarization. Nat. Nanotechnol. **5**, 297–301 (2010b)

S.H. Ko, Y.A. Song, S.J. Kim, M. Kim, J. Han, Nanofluidic preconcentration device in a straight microchannel using ion concentration polarization. Lab Chip **12**, 4472–4482 (2012)

S.J. Lee, D. Kim, Millisecond-order rapid micromixing with non-equilibrium electrokinetic phenomena. Microfluid. Nanofluid. **12**, 897–906 (2012)

T.A. Zangle, A. Mani, J.G. Santiago, Theory and experiments of concentration polarization and ion focusing at microchannel and nanochannel interfaces. Chem. Soc. Rev. **39**, 1014–1035 (2010)

Chapter 7
Conclusions and Future Prospects

RED is a promising method for capturing SGE in river estuaries without any moving parts. This book summarizes recent advancements in RED technology. For the successful application of RED, stack components and system design need to be enhanced for higher power density and lower cost. As commercial IEMs are not appropriate for RED systems, various tailored IEMs with high permselectivity and low resistance have been demonstrated at the lab-scale. In addition, nanochannels have emerged as an alternative approach; however, mass production of nanochannel membranes, which is required for their practical use, has not begun. In a flow channel, it is important to achieve uniform flow distribution, eliminate shadow effects, and lower the pressure drop. The profiled membranes have these properties. It is also crucial to determine the optimal operating conditions and develop maintenance technology to deal with fouling. The applications of RED are expanding as the power density of the technology improves, for example, brine recovery for desalination processes and waste heat recovery from industrial fields. However, despite recent achievements, full-scale RED plants are not yet available. Additional pilot-scale studies are required to validate the feasibility of RED.

© The Author(s), under exclusive license to Springer Nature Singapore Pte Ltd. 2019   45
D. Kim et al., *Energy Generation using Reverse Electrodialysis*,
SpringerBriefs in Applied Sciences and Technology,
https://doi.org/10.1007/978-981-13-0314-2_7

2